OUT OF THIS WORLD

Meet NASA Inventor Chris Walker and His Team's

Inflatable Stargazers

www.worldbook.com

World Book, Inc.
180 North LaSalle Street
Suite 900
Chicago, Illinois 60601
USA

For information about other World Book publications, visit our website at www.worldbook.com or call 1-800-WORLDBK (967-5325).

For information about sales to schools and libraries, call 1-800-975-3250 (United States), or 1-800-837-5365 (Canada).

© 2024 (print and e-book) by World Book, Inc. All rights reserved. No part of this publication may be reproduced, stored in a retrieval system, or transmitted in any form or by any means (electronic, mechanical, photocopying, recording, or otherwise) without written permission from World Book, Inc.

WORLD BOOK and the GLOBE DEVICE are registered trademarks or trademarks of World Book, Inc.

Produced in collaboration with the National Aeronautics and Space Administration (NASA).

Library of Congress Cataloging-in-Publication Data for this volume has been applied for.

Out of This World
ISBN: 978-0-7166-6564-9 (set, hc.)

Inflatable Stargazers
ISBN: 978-0-7166-6568-7 (hc.)

Also available as:
ISBN: 978-0-7166-6576-2 (e-book)
ISBN: 978-0-7166-6584-7 (soft cover)

Staff

Editorial

Vice President
Tom Evans

Senior Manager, New Content
Jeff De La Rosa

Writer
William D. Adams

Editor
Emma Flickinger

Curriculum Designer
Caroline Davidson

Proofreader
Nathalie Strassheim

Indexer
Nathaniel Lindstrom

Graphics and Design

Senior Visual
Communications Designer
Melanie Bender

Digital Asset Specialist
Rosalia Bledsoe

Acknowledgments

Cover	© Heather Roper, Northrup Grumman	22-23	© FreeFall Aerospace; Chris Walker
		24-25	ESO/B. Tafreshi
3	NASA; ESO/B. Tafreshi	27	NRAO; NRAO/AUI/NSF; The Ohio State University
4-5	NASA/WMAP Science Team	28-29	NASA
6-7	© leeborn/Shutterstock	30-31	© Heather Roper, Northrup Grumman
8-9	© Tony Baggett, Adobe Stock	33	The Ohio State University; Big Ear Radio Observatory and North American AstroPhysical Observatory
10-11	NASA		
12-13	NASA/ESA	34-35	NASA/JPL-Caltech
14-15	Chris Walker	37	© Look and Learn/Bridgeman Images
17	© Dorling Kindersley/Alamy Images	38-39	ESO
18-19	© Northrup Grumman	40-43	WORLD BOOK photo by Tom Evans
20-21	© FreeFall Aerospace	44	Chris Walker

Contents

- **4** Introduction
- **8** BIG IDEA: Balloon astronomy
- **10** STO and STO 2
- **12** Exploring with GUSTO
- **14** INVENTOR FEATURE: Scrapping with his brother
- **16** BIG IDEA: Inflatable reflectors
- **18** 10-Meter Suborbital Large Balloon Reflector
- **20** Steerable spherical reflector
- **22** Smallsat antenna
- **24** The birth of radio astronomy
- **26** INVENTOR FEATURE: Two degrees from Jansky
- **28** SALTUS: Giant leap on the cheap
- **30** SALTUS deployment
- **32** INVENTOR FEATURE: The Big Ear
- **34** SALTUS capabilities
- **36** INVENTOR FEATURE: Niépce brothers
- **38** An outer-space node for the EHT
- **40** INVENTOR FEATURE: Make your own radio
- **44** Astronomy power couple
- **45** Glossary
- **46** Review and reflect
- **48** Index

Glossary There is a glossary of terms on page 45. Terms defined in the glossary are in boldface type that **looks like this** on their first appearance on any spread (two facing pages).

Pronunciations (how to say words) are given in parentheses the first time some difficult words appear in the book. They look like this: pronunciation (pruh NUHN see AY shuhn).

Introduction

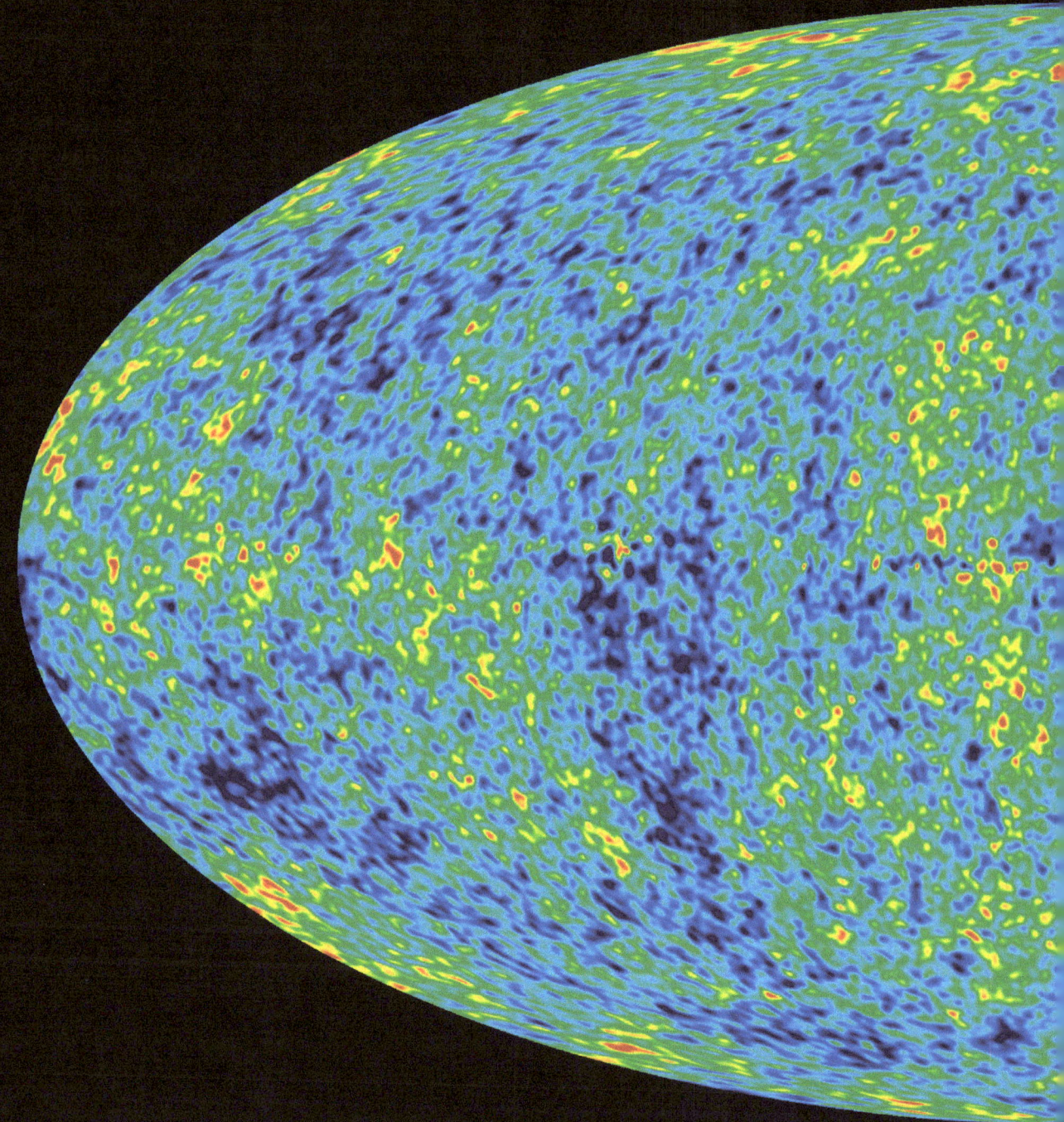

Visible light is just one small fraction of the **electromagnetic radiation** that fills our universe—like a single chapter in a larger book. To learn more about faraway objects—and even the origin of the universe itself—astronomers study electromagnetic

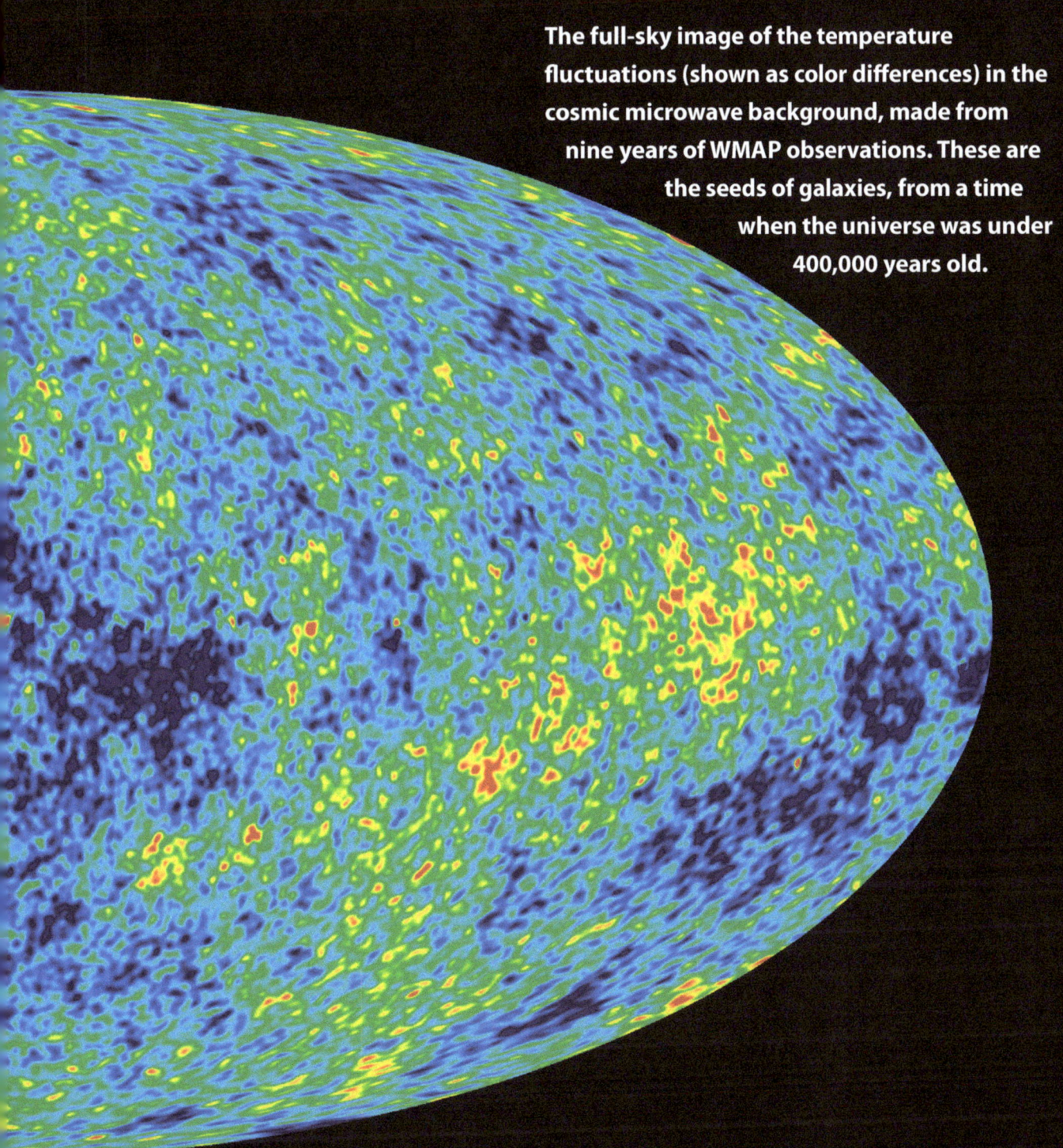

The full-sky image of the temperature fluctuations (shown as color differences) in the cosmic microwave background, made from nine years of WMAP observations. These are the seeds of galaxies, from a time when the universe was under 400,000 years old.

radiation with *wavelengths* longer than those of visible light. (Wavelength is the distance between peaks or crests of a wave, such as a light wave.) This longer wavelength **radiation** includes far-infrared, microwaves, and radio waves.

But, not all long-wavelength signals are detectable from Earth's surface. The water and oxygen **molecules** within our **atmosphere** block some of these long wavelengths. Also, people's use of such signals for communication drowns out part of the spectrum (overall range of wavelengths).

To study these longer wavelengths, astronomers must get their telescopes off the ground and above the atmosphere. The telescopes do not necessarily have to be launched into space. A high-altitude balloon can carry a telescope above 95 percent of the atmosphere, enabling it to capture valuable long-wavelength signals.

Chris Walker, an astronomer at the University of Arizona, wants to take the use of balloons even further—he wants the balloon to be part of the telescope itself. And, he has related ideas that could improve everything from cell phone reception to the study of black holes.

The NASA Innovative Advanced Concepts program. The titles in the *Out of This World* series feature projects that have won grant money from a group formed by the United States National Aeronautics and Space Administration, or NASA. The NASA Innovative Advanced Concepts program (NIAC) provides funding to teams working to develop bold new advances in space technology. You can visit NIAC's website at www.nasa.gov/niac.

Meet Chris Walker.

❝ I've been obsessed with radio signals and technology since I was five years old. Now, I'm coming up with new ways to observe long wavelengths of **electromagnetic radiation** in the universe—and to direct them here on Earth. ❞

Big idea: Balloon astronomy

Ground-based telescopes have been vital to our understanding of the universe. But, they cannot be used for all observations. Gases in Earth's **atmosphere** absorb some wavelengths of **electromagnetic radiation,** particularly those in the infrared and radio spectrum.

Astronomers can use orbital telescopes to make observations without the hindrance of the atmosphere. However, it is simply not possible to perform all observations this way. Rocket launches are extremely expensive. In addition, the light detection systems must be able to survive the harsh forces undergone during launch and in outer space. Once launched, orbital telescopes cannot easily be retrieved or repaired.

There is, however, a cheaper option. High-altitude balloons, such as those used by scientists who study the weather, can float to the edge of space, above 95 percent of the atmosphere. Astronomers can use these balloons to carry telescopes and other instruments. When the mission is finished, the instruments can be returned to Earth by parachute, possibly to be used again. Such a balloon mission costs a small fraction of a rocket launch.

> ❞ For the type of work I do, it's almost as good as being in space. ❞ —Chris

STO and STO 2

Chris is interested in **electromagnetic radiation** in the far-infrared portion of the spectrum. Far-infrared waves come from extremely cold matter, such as the clouds of dust and gas between stars. This material is called the *interstellar* medium. Water vapor absorbs far-infrared waves, however, and Earth's **atmosphere** is full of water vapor. Aside from an expensive rocket launch, the only way to study these signals is with balloon astronomy.

Chris oversaw NASA's Stratospheric Terahertz Observatory (STO) to better understand the interstellar medium. STO launched from Antarctica in 2012 using a high-altitude balloon. The success of the mission led NASA to schedule a follow-up. STO 2 took place in 2016, again launching from Antarctica, to study the formation of stars within the interstellar medium. It carried a telescope that had been in use since the 1990's.

> STO helped to demonstrate what you could do from a balloon in the far infrared. —Chris

STO-2 launch in Antarctica

Exploring with GUSTO

Chris and his team have built on the success of the STO missions to develop the Galactic/Extragalactic ULDB Spectroscopic Terahertz Observatory (GUSTO) mission. However, GUSTO will use a much more sophisticated instrument. It will simultaneously detect the far-infrared light emitted by carbon, nitrogen, and oxygen **atoms** from across our galaxy, the Milky Way, and another nearby galaxy, called the Large Magellanic Cloud.

Like the STO missions, GUSTO will fly from Antarctica using a long-duration balloon (LDB). But, this mission will stay in flight for a much longer time, between about 55 and 75 days.

Dust in the Large Magellanic Cloud, seen in this picture, glows in the far-infrared part of the spectrum.

❝ This GUSTO **payload** is huge. It's 25 feet [7.6 meters] wide and weighs about 2 tons. That dangles from a high-altitude balloon that flies at about 120,000 feet [37,000 meters]. The high-altitude balloon itself is about 140 meters [460 feet] in diameter at altitude—larger than a football field. ❞ —Chris

Inventor feature:
Scrapping with his brother

Chris grew up in a small town in North Carolina. There were seemingly boundless forests on the edge of town. Sometimes, people would dump things in the forest: old refrigerators, lawnmowers, and other junk.

Chris and his older brother would ride their bicycles into the forest and pick up the choice bits of junk. They would

A young Chris (center) with his mother and siblings

take them home and dismantle them.

> **Sometimes, you'd find an old tube radio that if you plugged it in, it would still work.** —Chris

They would turn the lights out and watch the faint glow of the **vacuum** tubes and listen to the distant voices coming through the speaker.

> **It was just magic. I was hooked from that point on.** —Chris

Radio technology became Chris's passion. He built crystal radios, got into ham radio, and eventually entered the field of **radio astronomy.**

Grade-school photograph of Chris

Big idea:
Inflatable reflectors

Serendipity means a beneficial chance occurrence—a happy accident. It sometimes plays a key role in scientific breakthroughs.

In 1982, Chris was in his apartment in Redondo Beach, California. He was making chocolate pudding on the stove when his mother called on the telephone. Chris covered the still-warm saucepan with plastic wrap. As he sat on the couch and talked to his mom, he absentmindedly set the covered saucepan on the floor next to him.

❝ About an hour later or so, I was watching TV and out of the corner of my eye I saw a light bulb floating over the arm of the couch!

"There was a little floor lamp off the edge of the couch, and the chocolate pudding saucepan with the plastic wrap was underneath. What had happened was the air between the chocolate pudding and the plastic wrap had cooled off, and in the process, sucked down the plastic wrap and formed a kind of mirror called a parabolic reflector.

"Light from the light bulb hit the plastic wrap, which is now parabolic in shape, and some fraction of that reflected up and formed a focus close to where I was looking ... producing an image of the light bulb. I saw this thing floating in midair!" —Chris

Chris was working on different projects at the time, but he was struck by the idea of flexible reflectors and vowed to come back to it.

10-Meter Suborbital Large Balloon Reflector

Decades later, Chris was able to put his serendipitous discovery to good use. He had already been involved with balloon astronomy. However, even the largest balloons can loft only modestly sized telescopes. Part of the reason is that the lenses and reflectors used are made of glass and metal and are therefore very heavy.

Balloons, of course, are very light. And the curvature of an inflated balloon resembles that of a traditional lens or reflector. Could a balloon be part of the telescope itself?

Chris sought to answer this question with the help of a NIAC grant. He developed the idea for the 10-Meter Suborbital Large Balloon Reflector. A 10-meter (33-foot) *spherical* (round) balloon would be mounted inside a much larger high-altitude balloon. Part of the smaller balloon would be coated with metallic paint to make it a reflector. Far-infrared waves would enter through both balloons and hit the reflector. The reflector would focus the waves on a receiver suspended inside the smaller balloon.

❚❚ The front of the balloon is clear, so far-infrared waves can pass on through. The back side of the sphere is metalized—like a fancy birthday balloon—and the waves reflect off that and come to a focus. ❚❚ —Chris

Steerable spherical reflector

A traditional reflector is a shape called a parabola, which focuses **radiation** on a single point. Balloons, on the other hand, tend to be spherical in shape. Spherical reflectors pose a unique challenge, but Chris and his team have turned it into an advantage.

❝ With a spherical reflector, as you use more and more of it, you get not a focal point but a focal line. So, you have to gather up all the waves along that focal line and collapse that focal line into a focal point that would then go into your detectors. ❞ —Chris

Chris invented a device called a phased array line feed that can concentrate a focal line. Furthermore, the phased array line feed can be digitally "steered." Multiple focal lines from the spherical reflector fall over the line feed. A computer or human operator can choose which parts of the line feed are active, thereby switching the feed. This ability enables the spherical reflector to focus on different regions of the sky—without any moving parts!

Down to Earth:

Ideas from space that could serve us on our planet.

The phased array line feed has potential outside the field of **radio astronomy.** It could help you check your social media posts! Cellular signals travel between a phone's antenna and special devices called *base stations*. Base stations are usually mounted on towers, but can be mounted directly on buildings in large cities. A company Chris co-founded called Freefall Aerospace has developed a base station that uses a phased array line feed. The FreeStar5G base station will deliver cellular data to phones in cities and other crowded areas. Even though the FreeStar5G is smaller than standard base stations, it is also more powerful. One FreeStar5G could replace up to four standard units!

Smallsat antenna

Inflatable balloons even have uses in space. Small satellites, also called *smallsats*, are a fast-growing field. The most famous type of smallsat is the CubeSat, which is made up of units based on a 10-centimeter (4-inch) cube. The standardization of CubeSats and the continued *miniaturization* (shrinking) of electronic components has enabled smallsat technology to soar.

One area where miniaturization has lagged, however, is in the ability to transfer data gathered from the smallsat to the ground.

❞ With a little tiny spacecraft, you don't have a big parabolic dish antenna that you can use to send the data down. ❞ —Chris

Chris's work on the large spherical reflector showed that inflatable balloons can be used to receive and focus signals. If a balloon can be used to receive signals, then it can be used to send them as well.

Chris and his team have designed a small balloon that can be used as a downlink antenna for a smallsat. The balloon and a small tank of *inflatant* take up less space and **mass** than a conventional antenna. And, the inflatable downlink can send far more data!

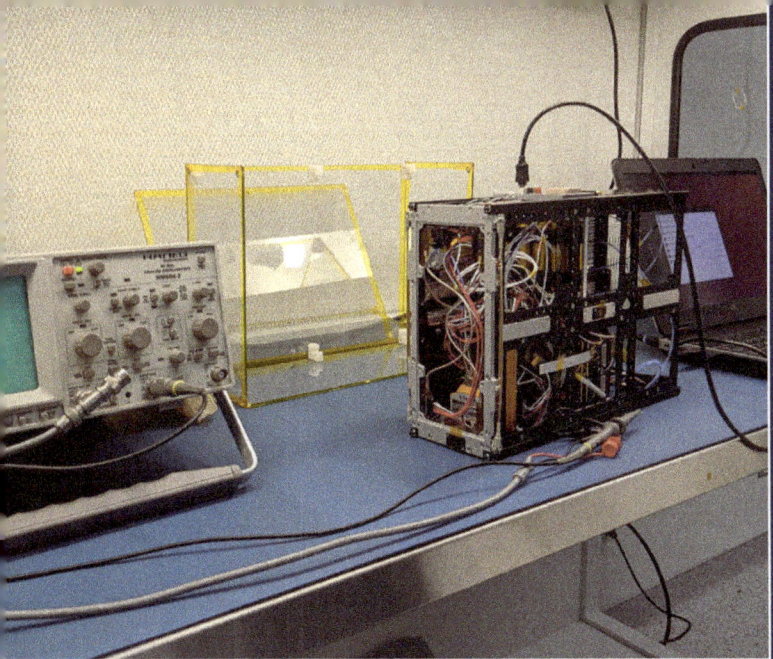
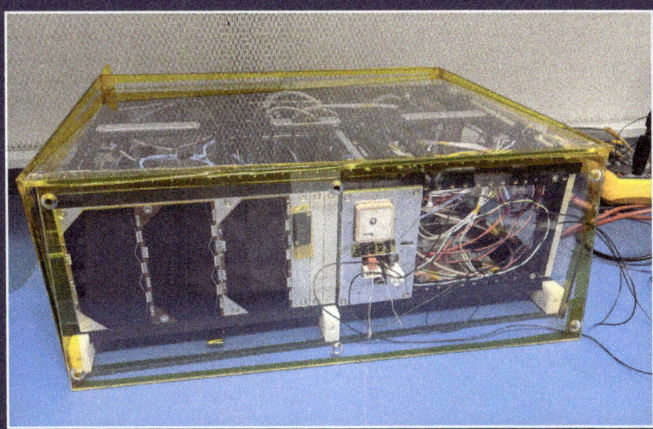

TOP LEFT: Testing the assembled CatSat. **BOTTOM LEFT:** The assembled CatSat packaged for delivery. **RIGHT:** Rendering of the deployed CatSat.

A high-altitude balloon test proved that such an inflatable reflector could rapidly send large amounts of photographic data back from space. The next step is to test the reflector in space. CatSat is an experimental CubeSat that will deploy and test such a balloon antenna from **orbit.** If successful, the inflatable antenna could revolutionize the design of smallsats and greatly expand their capabilities.

The birth of radio astronomy

A serendipitous discovery led to the development of the entire field of **radio astronomy.** For decades, Bell Labs was a research branch of what is now the telecommunications company AT&T. As NIAC does, Bell Labs encouraged scientists and **engineers** to try new things and work together to make great discoveries.

> The transistor was invented there, among many other things. In fact, radio astronomy, which is the kind of astronomy I do, was serendipitously discovered at Bell Labs. —Chris

Radio technology was developed in the late 1800's. It enabled people to communicate in real time overseas. However, nearby thunderstorms could produce static that interfered with the transmissions. In 1932, the Bell Labs physicist Karl Jansky set out to determine what time of day was best for such transmissions.

Jansky built a large dish that slowly swept the horizon in a 360-degree arc. He noticed that many unidentified radio emissions were coming from a certain part of the sky. After months of research, he realized that the center of our galaxy, the Milky Way, was emitting these radio waves.

Inventor feature:
Two degrees from Jansky

In 1934, John Kraus, a young physicist at Ohio State University (OSU), came to an astronomy conference. Kraus was excited to hear about Jansky's discovery of radio waves from the Milky Way.

> He went to listen to Karl Jansky back in 1934 at the very first presentation of **radio astronomy**… He realized how important this was, but no one else was in the room! —Chris

Kraus went back out into the conference and convinced more people to come in to listen to Jansky's talk.

Another early advocate for radio astronomy was an American inventor named Grote Reber. Reber read the work of Jansky and was inspired to build the first dedicated radio telescope in the backyard of his Wheaton, Illinois, home in 1937.

Top to bottom: Karl Jansky, Grote Reber, and John Kraus

Unlike NIAC, Bell Labs' support for new ideas was limited to those that its parent company could commercialize. Officials at Bell Labs figured there was no money to be made in radio astronomy, so they moved Jansky to other projects. Despite the dedication of Kraus and Reber, radio astronomy took many years to catch on.

But Kraus persevered, turning OSU into a major hub for radio astronomy as the field grew in importance. Chris was one of Kraus's Ph.D. students.

SALTUS:
Giant leap on the cheap

Balloon observatories work well for some forms of astronomy. But some observations require being entirely free of the **atmosphere** or being in a stable **orbit** around Earth or the sun. Chris and his team have developed a way to harness the advantages of inflatable technology in space. His Single Aperture Large-scale Telescope for Universe Studies (SALTUS) will be an infrared telescope with a 46-foot (14-meter) inflatable parabolic mirror. *Saltus* is Latin for *leap,* hinting at the enormous potential of the telescope.

❝ It'll be the next big thing after the James Webb Space Telescope (JWST). ❞ —Chris

SALTUS will not be the first large inflatable lens in space. The Inflatable Antenna Experiment (IAE) was an experimental telecommunications satellite with an inflatable receiver. It was successfully deployed from the space shuttle in 1996.

❝ What we're doing is taking what we've learned in the last 25 years in terms of implementing these inflatables and also using adaptive optics techniques to make such a telescope usable throughout the far-infrared up into the mid-infrared. ❞ —Chris

The space shuttle Endeavour deployed the Inflatable Antenna Experiment (IAE) and monitored its deployment. Although the experiment was a success, inflatables saw little use in space for the next 20 years.

The revolutionary JWST used complex folding mirror segments, making it extremely expensive. SALTUS will have fewer such delicate parts, making it much cheaper than Webb.

SALTUS deployment

SALTUS will launch aboard a conventional rocket and move into an **orbit** around the sun, like JWST. Once there, a boom carrying the stowed reflector will deploy, putting it the correct distance from the detection system. The 46-foot (14-meter) reflector will then inflate. Then, a sunshield will deploy to allow the reflector to cool to about 25 degrees above absolute zero. Absolute zero is temperature at which **atoms** and **molecules** have the least amount of heat possible. A cold reflector lowers the the thermal background noise seen by the detectors, making the telescope more sensitive to the feeble far-infrared **electromagnetic radiation** from the distant universe.

Micrometeoroids zing around the **solar system.** They burn up in Earth's **atmosphere** but pose a serious hazard in space. The SALTUS mirror will be thinner than a party balloon. Would a single micrometeoroid impact reduce the reflector to shiny confetti?

Chris and his team have performed tests suggesting that the micrometeorites would only produce tiny holes in the reflector, rather than large gashes.

Because space is a **vacuum,** the reflector does not need much inflatant. To compensate for these unavoidable leaks, SALTUS will carry extra inflatant. Just a small tank will be enough to keep the reflector inflated for a mission lasting many years.

" It only takes a wisp of inflatant to inflate the reflector. So, if you prick it, it doesn't pop, gas just slowly leaks out through the tiny hole. " —Chris

Inventor feature:
The Big Ear

During Chris's time at Ohio State University, he took part in the search for extraterrestrial intelligence (SETI). He worked at the Ohio State University Radio Observatory, which was designed by John Kraus. Chris spearheaded an overhaul of the device's listening antenna. Because the observatory was primarily used for listening for signs of extraterrestrial intelligence, it was nicknamed the Big Ear.

In 1977, the Big Ear recorded a powerful, 72-second radio beam from the constellation Sagittarius. American astronomer Jerry Ehman noticed the anomaly in the data printout, circled it, and wrote "Wow!" next to it. Photocopies of Ehrman's excited scrawl appeared alongside newspaper stories of the discovery, leading it to be nicknamed the "Wow!" signal.

The "Wow!" signal was a narrow band of radio waves that astronomers would not expect to be

Big Ear and "Wow!" signal printout

produced by a star or other planetary body. It could have been produced by technology, a possible sign of intelligent life elsewhere in the universe. But, no similar signal has been detected from that field of sky despite decades of listening.

The Big Ear was demolished in 1998. But, other observatories are taking up its search. Projects are underway to meticulously scan each star in the area of sky from the which the signal could have originated.

SALTUS capabilities

Think back to the STO and GUSTO projects. Because water vapor in our **atmosphere** blocks and reflects infrared waves, Chris used high-altitude balloons to gather measurements. All life as we know it requires water. Therefore, infrared astronomy is perfectly suited to observing potentially habitable **exoplanets.** If their atmospheres contain water vapor, it will interact with infrared waves in ways that SALTUS can observe.

❞ For the SALTUS mission, there are two main science objectives. The first one is to understand the formation of planetary systems and where the **habitable zone** within the new planetary system would be. That would be the region within that protoplanetary system where you could have water on the surface for life. ❞ —Chris

SALTUS will sniff out water vapor and other **molecules** useful to life in promising locations closer to home, too. Europa, a moon of Jupiter, and Enceladus, a moon of Saturn, both have surfaces made of ice. Scientists suspect that both have oceans of liquid water beneath their icy crusts.

❞ We can also look at the transport of water through the solar system, from **comets** to planetary bodies to the formation of Earth's oceans. ❞ —Chris

Astronomers could use SALTUS to determine if the water on Earth has been here since it was formed or was delivered later by comets.

Artist's impression of a view from an exoplanet in the TRAPPIST-1 system, where seven planets orbit a small star in or near the habitable zone

Inventor feature:
Niépce brothers

Chris has been drawn to scientists and inventors who were ahead of their time.

❙❙ For most leading-edge science and leading-edge technologies, it takes a while to convince people that they're doable. The more game-changing your technical approach or your science is, the more convincing you have to be…You can see examples of this throughout history. ❙❙ —Chris

Two such visionaries were the brothers Claude and Nicéphore Niépce. They invented the first operating internal combustion engine, called the *pyreolophore,* which was patented in 1807. Napoleon Bonaparte, who reigned as emperor of France at the time, signed the patent. The brothers spent decades trying to commercialize their invention in France and England, to no avail.

In this illustration, Nicéphore Niépce (left) and Louis J. M. Daguerre (right) examine an early photograph they created in their laboratory.

While Claude spent the family fortune looking for investors for the *pyreolophore,* Nicéphore experimented with using *photosensitive chemicals* (chemicals that react with light) to capture images on metal plates. He called his technique *heliography*. In 1826, he took what is widely regarded as the first photograph. One of his *protégés* (followers), Louis J. M. Daguerre, went on to formulate improvements to the technique. He created the first commercially successful photography process, called *daguerreotyping*.

An outer-space node for the EHT

The Event Horizon Telescope (EHT) is a global network of ground-based telescopes established to produce images of black hole **event horizons.** Because the telescopes are spread throughout the globe, the EHT essentially acts as a single telescope with a diameter the size of Earth. That's huge, but black holes are incredibly far away and difficult to image. Even this network of telescopes syncing its observations with pinpoint accuracy and using ingenious computer software can only create blurry, low-resolution images of black holes.

❝ What we could do with a SALTUS-like telescope is add a space node to the EHT. ❞ —Chris

With a SALTUS-like telescope, the effective diameter of the EHT will be far larger than that of Earth. Not only would a SALTUS-equipped EHT take higher-resolution images of black holes, but it would also be able to make videos. We could watch plumes of gas—and even stars—fly around a black hole's event horizon.

❝ We can use SALTUS to look for the most distant galaxies and understand what their content is and how they evolved over time. ❞ —Chris

With a SALTUS-like telescope, this image of SgrA* could have much better resolution.

Because light from the farthest parts of the universe takes billions of years to reach us, far-away objects are time capsules of the universe. With SALTUS, astronomers will be better able to characterize galaxy formation and evolution.

Beyond NIAC

NASA is considering SALTUS for a **probe**-class mission. The technology has been developed, so it is ready to be built and launched! This mission is so close to going to space because of NIAC's early investment in Chris's 10-meter Spherical Large Balloon Reflector.

❞ Why NIAC is so cool is that it's really, really hard to get funding for game-changing ideas. NIAC can help provide that initial influx of funding that's required to get what initially seems like these out-of-the-box ideas into reality, enough that either they can be picked up by a large NASA program or commercialized. ❞ —Chris

39

Inventor feature:
Make your own radio

When Chris was a student, he performed many experiments featured in the pages of *The World Book Encyclopedia,* including using a light beam for communication and building his own radio. Here is a step-by-step guide on how to build your own radio that Chris has prepared for you!

What you will need:

- **Cardboard tube**
- **Thin, insulated wire**
- **Germanium diode**
- **Crystal radio earphone**
- **Screw, washer, and nut**
- **Copper strip, about 6 inches (15 centimeters) long**
- **Push pins**
- **Cardboard or foamboard square, about 12 x 12 inches (30 x 30 centimeters)**
- **Sandpaper**
- **Sharp scissors or wire cutters**
- **Clear tape**

Directions

1 Ask an adult to drill a hole at one end of the copper strip large enough for the screw to fit through.

2 Poke a small hole about ½ inch (1 centimeter) from each end of the cardboard tube. Put one end of the wire through one of the holes and pull about a foot (30 centimeters) of it through the hole. Tape the wire to the edge of the tube with a small strip of clear tape.

3 Ask a friend or an adult to hold the roll of wire. Hold the cardboard tube and wind the wire onto it. Each coil should nestle next to the one before it. Keep adding coils until you reach the hole at the other side of the tube. Then, unspool another foot (30 centimeters) of wire and cut it off near the spool.

4 Thread the extra wire through the hole and tape it to the edge of the cardboard tube as you did in step 2. The tube wound with wire is the tuning coil that will allow you to change radio stations.

5 Poke a hole in the cardboard or foamboard base near the middle of one edge. Thread the washer and copper plate onto the screw and insert the screw through the hole in the base. Thread the nut onto the screw, fastening it loosely.

Inventor feature:
Make your own radio cont.

6 Place the tuning coil on the base so that the copper strip can touch the top of the wire coils. Secure each side of the roll to the base with a push pin.

7 Loop one wire from the diode around a push pin. Loop one wire from the earphone around the same push pin.

8 Loop the other wire from the diode around the screw holding the copper plate. Affix the push pin to the base.

9 Unspool some wire and connect it to a grounded metal object, such as plumbing pipes. Run the wire to where the radio is being assembled and cut it from the spool.

10 Run a very long piece of wire (about 30 feet or 10 meters) to use as the antenna. The antenna will provide better reception if it leads outdoors and is attached to a tree or pole—ask an adult for help. Run the antenna wire to where the radio is being assembled and cut it from the spool.

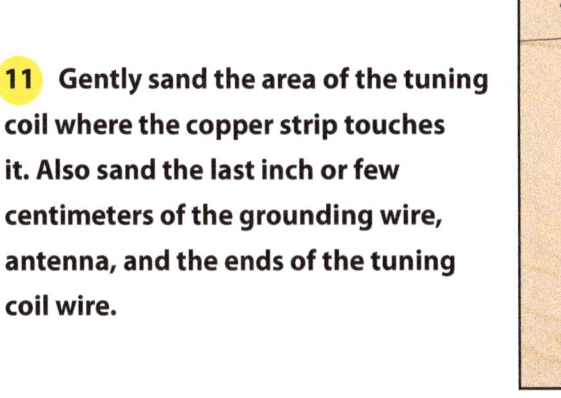

11 Gently sand the area of the tuning coil where the copper strip touches it. Also sand the last inch or few centimeters of the grounding wire, antenna, and the ends of the tuning coil wire.

12 Loop one end of the tuning coil and the end of the antenna wire around a push pin. Loop the other end of the tuning coil wire, the grounding wire, and the other wire from the earphone around a second push pin. Affix both push pins to the base.

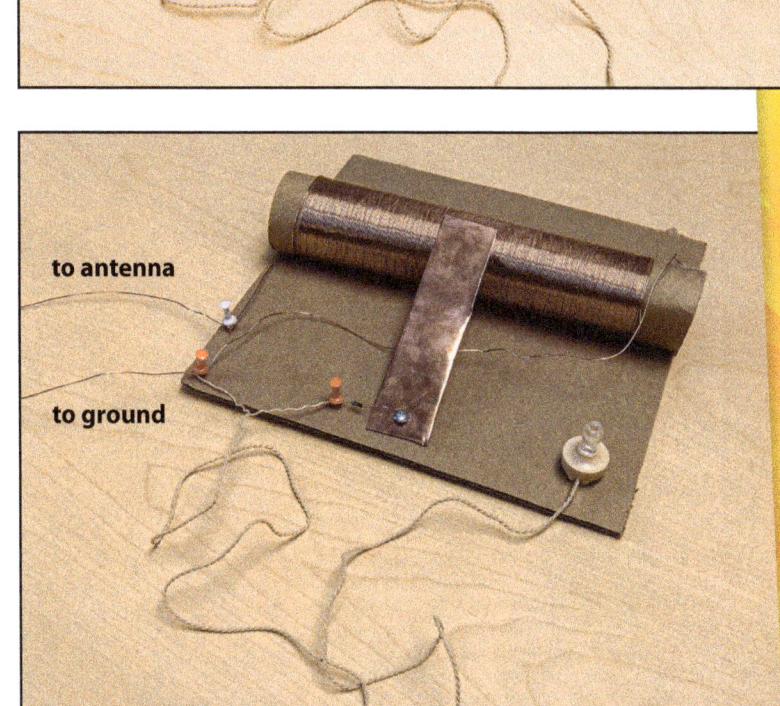

13 Position the copper strip on the tuning coil, tightening the nut and screw assembly as necessary to make a good connection.

14 Bring the earphone to your ear. You should pick up static or radio stations through the earphone. Move the copper strip to different points along the length of the tuning coil to pick up different radio stations.

> ❞ If you can't hear anything, double-check your wire connections. Don't despair if it doesn't work the first time—just keep trying! ❞ —Chris

Astronomy power couple

Chris's wife, Connie Walker, is also an astronomer. She is active in astronomy outreach and raising awareness of **light pollution.** She also advocates against the effects of huge satellite constellations that are now being launched. Though such satellites promise unprecedented connectivity, their sheer number poses a major threat to ground-based and balloon astronomy.

Glossary

atmosphere the mass of gases that surrounds a planet.

atom one of the most basic units of matter, consisting of a *nucleus* (core) of particles called *protons* and *neutrons* with tiny particles called *electrons* moving around the nucleus.

comet an icy body that releases gas or dust.

electromagnetic radiation energy given off in the form of *oscillating* (moving back and forth) electric and magnetic fields. Visible light, infrared rays, ultraviolet light, and X rays are all kinds of electromagnetic radiation.

engineer a person who uses scientific principles to design structures, such as bridges and skyscrapers, machines, and all sorts of products.

event horizon the surface of a black hole. At the event horizon, the pull of gravity becomes so strong that nothing known can escape it.

exoplanet (extrasolar planet) a planet that orbits a star other than the sun.

interstellar between the stars.

light pollution the effects of artificial lighting that harm the environment or interfere with people's view of the night sky.

mass the amount of matter something contains.

micrometeoroid an object less than 0.04 inch (1 millimeter) wide floating in space.

molecule two or more atoms bonded together.

orbit a looping path around an object in space; the condition of circling a massive object in space under the influence of the object's gravity.

payload the useful load carried by a vehicle.

probe a rocket, satellite, or other uncrewed spacecraft carrying scientific instruments, to record or report back information about space.

radiation energy given off in the form of waves or tiny particles of matter.

radio astronomy a branch of astronomy dealing with the study of objects in space by means of radio waves that these objects give off.

solar system the sun and everything that travels around it, including Earth and all the other planets and their moons.

vacuum empty space without even air in it.

Review and reflect

Now that you've finished reading about Chris Walker, use these pages to think about his experiences and telescope technologies in new ways. As you work, reflect on the importance of creative problem solving, curiosity, and open-mindedness in life.

Complex problems and creative solutions

Why are astronomers interested in studying long-wave electromagnetic radiation?

What are some of the problems associated with studying long-wave radiation?

How does Chris Walker hope to overcome these challenges with the 10-Meter Suborbital Large Balloon Reflector? What makes this solution so innovative?

Visit www.worldbook.com/resources to download sample answers, blank graphic organizers, and a rubric to evaluate writing.

Inspiration can come from anywhere!

Use a graphic organizer like the one below to map out your ideas. What ideas or experiences led to Chris's innovative solution?

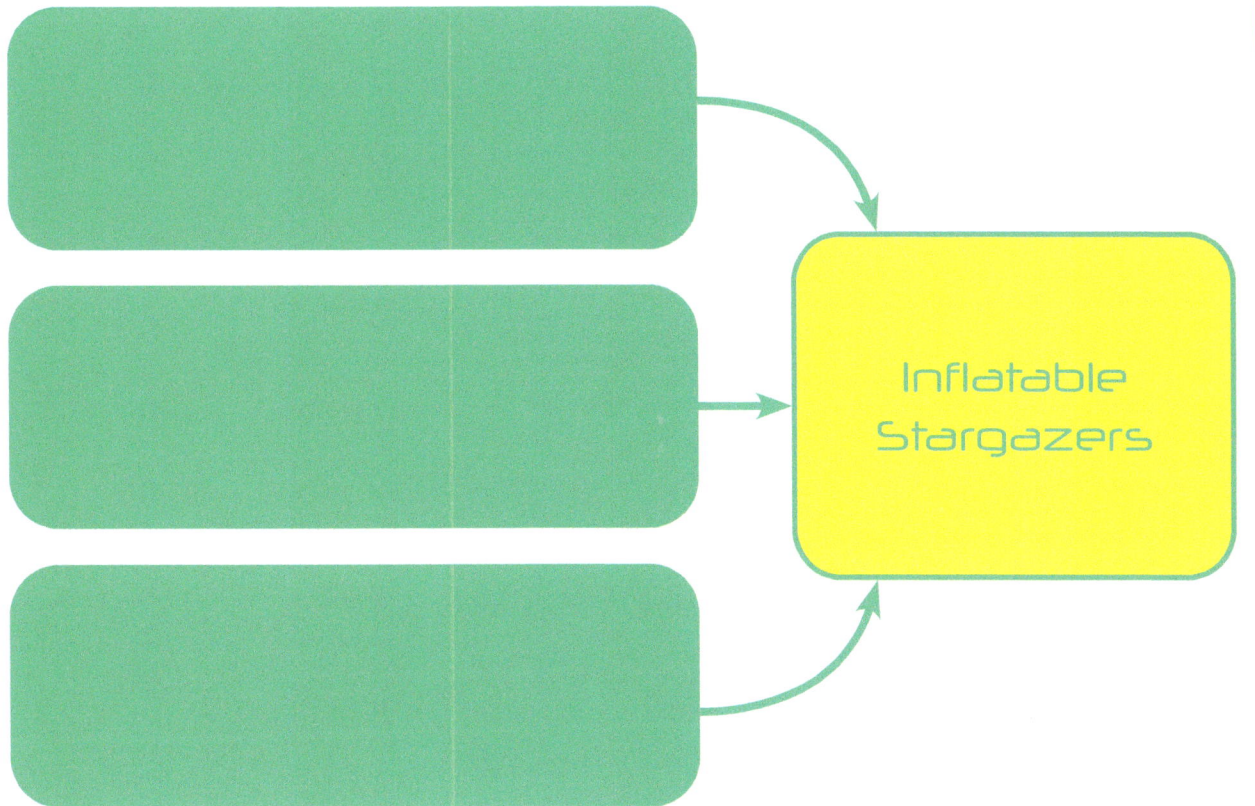

Write about it!

Think about Chris's experiences in life and as a NIAC Fellow.

How have Chris and his team used unconventional ideas to develop new telescope technologies? How can thinking outside the box help develop innovative solutions?

Index

A
absolute zero, 30
Antarctica, 10-12
atmosphere, 6, 8-10, 28, 30, 34
atoms, 12, 30

B
base stations, 21
Bell Labs, 24, 27
Big Ear, The, 32-33
black holes, 6, 38-39

C
carbon, 12
CatSat, 22-23
comets, 34
cosmic microwave background, 4-5
CubeSats, 22-23

E
electromagnetic radiation, 4-5, 7-8, 10, 30
Enceladus, 34
Endeavour (space shuttle), 28-29
Europa, 34
Event Horizon Telescope (EHT), 38-39
event horizons, 38
exoplanets, 34-35

F
far-infrared waves, 5, 10, 12-13, 18-19, 30

G
Galactic/Extragalactic ULDB Spectroscopic Terahertz Observatory (GUSTO) mission, 12-13, 34

H
habitable zones, 34-35
high-altitude balloons, 6, 10, 13, 18, 23, 34

I
ice, 34
Inflatable Antenna Experiment (IAE), 28-29
inflatants, 22, 30-31
internal combustion engines, 36
interstellar medium, 10

J
James Webb Space Telescope (JWST), 28-30
Jansky, Karl, 24, 26-27
Jupiter, 34

K
Kraus, John, 26-27, 32

L
Large Magellanic Cloud, 12-13
light pollution, 44

M
micrometeoroids, 30
microwaves, 4-5
Milky Way galaxy, 12, 24, 26
mirrors, 17, 28-30
molecules, 6, 30, 34

N
NASA Innovative Advanced Concepts program (NIAC), 7, 18, 24, 39
National Aeronautics and Space Administration (NASA), 7, 10, 39
nitrogen, 12

O
oceans, 34
orbital telescopes, 8-9
orbits, 23, 28, 30, 35
oxygen, 6, 12

P
phased array line feeds, 20-21
photography, 37
planetary systems, 34-35

R
radio astronomy, 15, 24, 26-27
radio waves, 5, 24, 26, 32-33
Reber, Grote, 26-27
reflectors, 16-20, 22-23, 30-31, 39
rockets, 9-10, 30

S
satellites, 22-23, 28-29, 44
Saturn, 34
Single Aperture Large-scale Telescope for Universe Studies (SALTUS), 28-30, 34, 38-39
smallsats, 22-23
solar system, 30, 34
Stratospheric Terahertz Observatory (STO), 10-12, 34

T
10-Meter Suborbital Large Balloon Reflector, 18, 39
TRAPPIST-1 (planetary system), 34-35

V
vacuum, 15, 30
visible light, 4-5

W
water, 34
water vapor, 6, 10, 34
wavelengths, 4-8
"Wow!" signal, 32-33

www.ingramcontent.com/pod-product-compliance
Lightning Source LLC
Chambersburg PA
CBHW060933170426
43194CB00023B/2954